Ogi Orange
THE BIG RACE

ISBN 979-8-9875895-0-2

Published by Pybabi Publishing

This book is dedicated to my dear wife Samantha whose unwavering support has been my inspiration.

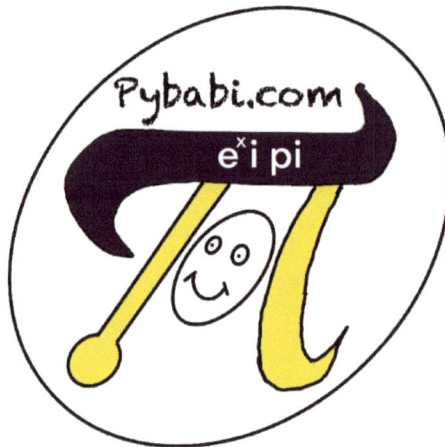

Message for instructor, teacher, or parent:

This book is best used by allowing one of the children who can read to lead the group after you have done it first as a model for them to imitate. Then they should read the story or section, ask the questions in the MATCH and ANSWER sections to their fellow students or if in a family to their brothers and sisters. As they lead the group you can monitor their progress. As the monitor you should answer any questions that arise naturally from the discussion. After one child has taken the lead, the role of teacher should be rotated to another child and then repeated until each child has successfully taught the material before advancing to another book. Since they are learning to teach the material they will naturally pay sufficient attention to learn the material and the lessons will stick with them. As the monitor, it may be best to intervene only as necessary allowing the children to solve and figure things out on their own until they need assistance.

If the child cannot read the parent can read and explain the lesson being taught. As soon as the child can understand even the simplest principle or idea have them to explain it back to you as if you are the student and they are the teacher.

When children feel they have an important role in education they will naturally learn what is necessary to play that role.

Of course this is not the only way to use the material, as you help the student you will soon discover which methods are best depending on the age and ability of the child. Please note that the illustrations in the book are specifically designed not to be perfect. They are only the beginning of a bigger idea or concept that will later be explained in greater detail. By keeping this in mind you will not have to push the student into trying to understand everything at the same time.

Thank you for choosing to use this material.

pybabi.com

OGI ORANGE

THE BIG RACE

Part I

Written and Illustrated by Lee Chau

OGI ORANGE

THE BIG RACE

Part II

Written and Illustrated by Lee Chau

OGI ORANGE

THE BIG RACE

Part III

Written and Illustrated by Lee Chau

OGI ORANGE

THE BIG RACE

Part IV

Written and Illustrated by Lee Chau

This is my book.

Date: _____

Name:

_____

I love you Space

Please sign, date and leave words of encouragement

I love you Space

Please sign, date and leave words of encouragement

.

I love you Space

Please sign, date and leave words of encouragement

Hi, I am Ogi Orange! I am the fastest fruit in the valley!

Mr Lary Berry, am I the fastest fruit in the valley?

I thought Banana Balii was the fastest fruit in the valley.

2

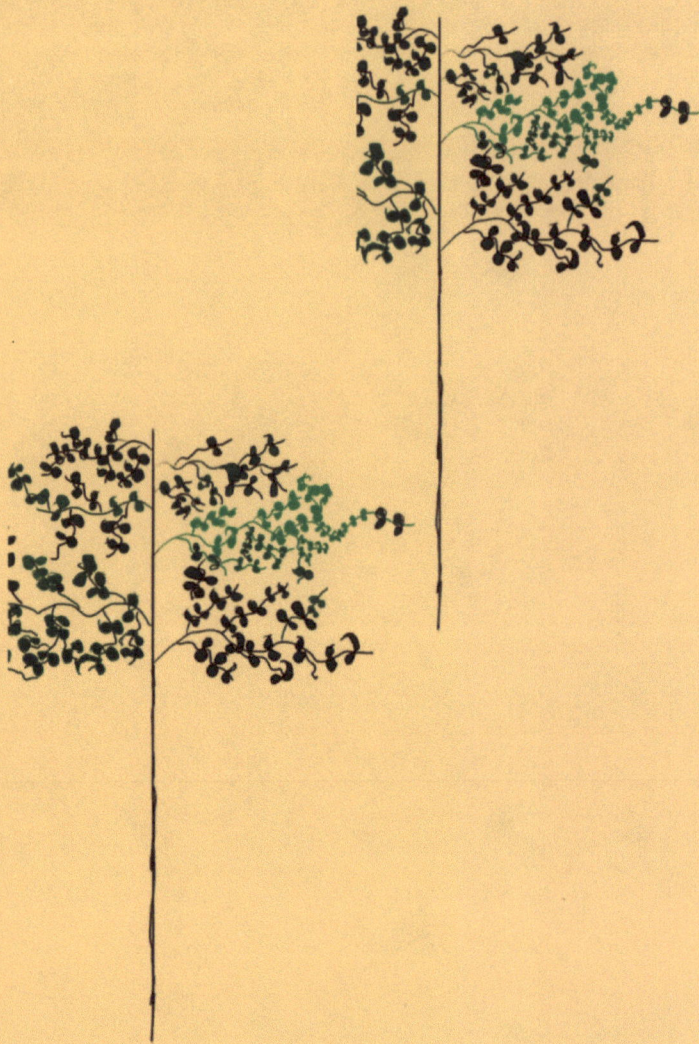

I must find this
Banana Balii. Where is
he? Have you seen him?

5

I think someone is looking for me. Do you know who? 6

Hey Banana Balii are you the fastest fruit in the valley?

Hi there, I am Mayor Pear. I think a race is in the air.

Of course I am! Just ask Wally or Sally I am Banana Balii the fastest fruit in the valley!

8

10 Long Track

5 Long Track

9

Ogi Orange has to run 10 Long. Balii Banana only has to run 5 Long. Who do you think will win?

10

10 Long Track

5 Long Track

11

Who is the fastest fruit in the valley? Ready, Set, Go!

12

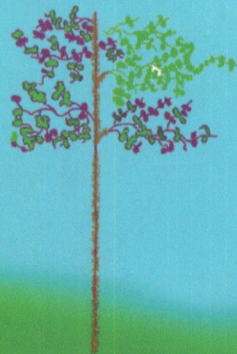

They are both running fast! Wow look at them go!

Go Balii go!

14

15

They are both halfway to their finish lines!

halfway

16

Congratulations! You both finished the race!

Finished

18

Fastest?

So, how do we find out who was the fastest?

We have to compare your AVERAGE speed. I will compare Ogi Orange's speed to Bali Banana's speed. But in order to truly understand speed, you have go to Castle Calculus and defeat 'Bender Pretender' if you can defeat him you will know 'The Great Secret of Speed!

Hi I am SR Star whether you are near or far I will be wherever you are. You will need me to be able to defeat Bender Pretender.

What is average speed?

Let's find our average speed to see 'who is the fastest fruit in the valley?'

Ok, but I must get The Secret to Speed! After we find out the average speed I have to defeat Bender Pretender!

22

MATCH 1

What type of fruit is Balii?

What type of fruit is Ogi?

What type of fruit is Irene?

What type of fruit is the mayor?

a tangerine

a pear

an orange

a banana

MATCH 2

Did they both finish the race at the same time?

How far did Balii run?

What words can we use to replace the word 'Long'?

How far did Ogi run?

miles, meters or kilometers

10 Long

Yes

5 Long

MATCH 3

When Ogi was passing Balii's finish line, how far had he run to his finish line?

What time did they both finish?

Halfway

Did Ogi and Balii understand who was the fastest?

What do Mayor Pear need to do in order to find out who ran the fastest?

6 Minutes

no

Compare the average speed of each runner.

MATCH 4

When we say 'Speed' in everyday life. What are we talking about?

Is average speed and average velocity the same thing?

Average Speed

Is speed 'at one point' and 'average speed' the same thing?

Will Mayor Pear compare their speed at one point or their average speed?

Average Velocity comes from Average Speed

no

Average Speed

MATCH 5

When you are riding in the car and you ask 'what is our speed?' What will the driver try to tell you?

Where must Ogi go to find 'The Secret of Speed'?

The 'average speed'

Who must Ogi defeat in order to get The Secret of Speed?

Who does Ogi need to help him to defeat Bender Pretender?

Castle Calculus

Bender Pretender

SR Star

ANSWER 1

What type of fruit is Balii?

What type of fruit is Ogi?

What type of fruit is Irene?

What type of fruit is the mayor?

ANSWER 2

How far did Balii run?

Did they both finish the race at the same time?

What words can we use to replace the word 'Long'?

How far did Ogi run?

ANSWER 3

When Ogi was passing Balii's finish line, how far had he run to his finish line?

What do Mayor Pear need to do in order to find out who ran the fastest?

Did Ogi and Balii understand who was the fastest?

What time did they both finish?

ANSWER 4

Will Mayor Pear compare their speed at one point or their average speed?

When we say 'Speed' in everyday life. What are we talking about?

Is speed 'at one point' and 'average speed' the same thing?

Is average speed and average velocity the same thing?

ANSWER 5

When you are riding in the car and you ask 'what is our speed?' What will the driver try to tell you?

Who must Ogi defeat in order to get The Secret of Speed?

Who does Ogi need to help him to defeat Bender Pretender?

Where must Ogi go to find 'The Secret of Speed'?

Teachers Space

Friends Space

Please sign, date and leave words of encouragement

Mess up Space

Do whatever you want here.

Part I

Written and Illustrated by Lee Chau

Part II

Written and Illustrated by Lee Chau

Part III

Written and Illustrated by Lee Chau

Part IV

Written and Illustrated by Lee Chau

Let's go to part II

Can you share your story?

Tiktok @ogi.orange

Twitter @ogi_orange

Instagram @ogi.orange

Email: ogiorange@pybabi.com

www.ingramcontent.com/pod-product-compliance
Lightning Source LLC
Chambersburg PA
CBHW041546040426

42447CB00002B/67